YOUR KNOWLEDGE HAS VALUE

- We will publish your bachelor's and master's thesis, essays and papers

- Your own eBook and book - sold worldwide in all relevant shops

- Earn money with each sale

Upload your text at www.GRIN.com and publish for free

Akampurira Abraham

Development Theory, Policy and Planning

Development Studies

GRIN Verlag

Bibliografische Information der Deutschen Nationalbibliothek:

Die Deutsche Bibliothek verzeichnet diese Publikation in der Deutschen National-
bibliografie; detaillierte bibliografische Daten sind im Internet über http://dnb.d-
nb.de/ abrufbar.

Imprint:

Copyright © 2011 GRIN Verlag GmbH
Druck und Bindung: Books on Demand GmbH, Norderstedt Germany
ISBN: 978-3-656-34997-6

This book at GRIN:

http://www.grin.com/en/e-book/207232/development-theory-policy-and-planning

ATLANTIC INTERNATIONAL UNIVERSITY.

NAME: AKAMPURIRA ABRAHAM :

COURSE: DEVELOPMENT THEORY, POLICY AND PLANNING.

MAJOR. REGIONAL DEVELOPMENT.

DATE: 05-12 - 2011

CONTENTS.   **Page**

Acronyms

AIDS- ACQUIRED IMMUNE DEFICIENCY SYNDROME.

ADB- AFRICAN DEVELOPMENT BANK.

CBO -   COMMUNITY BASED ORGANISATION.

CSO-   CIVIL SOCIETY ORGANISATION.

GDP-   GROSS DOMESTIC PRODUCT.

GNP-   GROSS NATIONAL PRODUCT.

IMF-   INTERNATIONAL MONETARY FUND.

MDG'S- MILLENIUM DEVELOPMENT GOALS.

NEMA-   ENVIRONMENTAL MANAGEMENT AUTHORITY.

NGO-   NON GOVERNMENTAL ORGANISATION.

PEAP- POVERTY ERADICATION ACTION PLAN.

PEAP- POVERTY ERADICATION ACTION PLAN.

UIA-   UGANDA INVESTMENT AUTHORITY.

UNDP- UNITED NATIONS DEVELOPMENT PROGRAM.

UPE-   UNIVERSAL PRIMARY EDUCATION.

URA-   UGANDA REVENUE AUTHORITY.

USE-   UGANDA SECONDARY EDUCATION.

UWA-   UGANDA WILD LIFE ASSOCIATION.

WTO-   WORLD TRADE ASSOCIATION.

Chapter 1

1.1 introduction.

It is the wish of every community and every nation to move from one stage of development to another. Development is holistic approach that involves structural institutional changes and social economic transformation, in addition to increased outputs and incomes. Development also encompasses change in peoples' customs and beliefs that are a hindrance to development programs. Economic growth involves urbanization, industrialization and increased and appropriate use of technology in all sectors of the economy. The government major role is to provide good welfare to its citizens. It is also obliged to facilitate economic players to actively participate in economic activity through provision of infrastructure. All this is done through the process of development theory, policy and planning.

Developing countries have an outstanding characteristic of high poverty levels with low national income and per capita incomes. Proper policy planning is meant to increase the national incomes of these countries. There is low savings as a result of low incomes and this has a negative effect on capital accumulation. Thus the end result has been a viscous cycle of poverty where there is low savings, low investments, low capital accumulation, and low productivity, low quality of life where there is generally misery, lack of freedom and dignity. For developing countries to come out of this precarious state, proper policy and planning must be adopted. This paper will cover the aspects of development theory, policy and planning because for sustainable development these aspects are a paramount.

1.2. Purpose of the study.

The course will enable students to;

- Identify the objectives of development planning.

- Identify the impact of policy in achieving sustainable economic growth and development.

Chapter 2

2.1 What is the theory about?

A theory is a simplified representation of reality. This can be; social, economic and political realities. Institutionally these realities have; community elements; which include the social political and economic changes at this level. Markets/business level which usually show the economic changes. The political/management of society which usually shows the political changes in terms of good and bad governance.

Theory can appear in different levels. These include.

- Conceptual theory; this shows the dimensions of the social, economic reality.
- Explanatory theories. This explains the causes and effects. For example when reducing poverty you must have an explanation to why it is rampant. For instance the trickle down theory of development
- Predictive theory. This explains what will happen in the future; given the current situation.
- The policy theory. This level refers to the policy instruments used given the current development theory.

Theory falls in any of the following categories;
- Social economic theory.

This can be a theory of social economic change. For instance the theory of poverty reduction or theory of political change. These are theories that show the relationship between social economic change and political change. This theory tries to explain the interaction between the economy and politics

- Theory of public policy making.

This is an explanation of how a policy is made or determined. It questions the key determinants of resource allocation in agriculture as well as infrastructure development among other sectors of the economy. This is more done by government donors or by the community. It may also include influence of politicians in cabinet and politicians' parliament, the influence of the president as well as influence of civil society organizations (CSO). This theory is how the policy agenda is determined. It implies a question "Can poverty be reduced without the poor influencing the policy agenda to eradicate poverty?"

What is the relevance of approaches to policy making

There are some theories that have been driving the policy agenda in Uganda and East African countries which include among others;

- The structural adjustment theory which has been driving the economic policy management
- The basic needs theory which has been driving both the economic and social policy management.
- The human rights theory which also drives both economic and social development policy management.

2.2 DEVELOPMENT THEORY.

Many scholars have put forward their ideas behind the forces for and against development. These schools of thought have some differences and similarities. They have however become the basis of planning for economies to achieve development. Four schools of thought dominate the literature on economic development after the Second World War when economies have been undergoing a restructuring program to overcome most of the economic setbacks. These theories include;

- The linear stages of growth model.
- Theories and patterns of structural change.
- The international dependence revolution.
- The neo-classical, free market counterrevolution.

2.2.1 The linear stages model

In this theory, a right combination of savings, investment and foreign aid were the only necessary parameters to bring about economic growth. According to this theory, development rightly meant rapid aggregate economic growth. These were theorists of the 1950's and early 1960's.

2.2.2 Theories and patters of structural change.

This was an improvement over the linear stages model. It emphasizes structural sustenance of economic growth. It entails the structural process an economy goes through to attain economic growth. Underdeveloped economies focus on transforming their domestic structures from heavy reliance on traditional systems of subsistence agriculture to modern industrialized and service sectors. Tools of neoclassical price and allocation theory and modern econometrics are necessary for economies transformation.

2.2.3 The international dependence revolution.

According to this theory economic growth should be attained through a political orientation. It looked at underdevelopment in terms of institutional and structural economic rigidities, proliferation of dual economies and dual societies, and international and domestic power relations across nations. Dependence theorists put more emphasis on external and internal institutional and political effects on economic growth.

It is noted that dependent theorists offer an appealing explanation of why poor countries have remained underdeveloped but do not give formal or informal techniques of how these countries can initiate and sustain development. Secondary it looks as if countries should be in autarky, restriction of interdependence of economies and only trading with the counterpart developing economies. This has been disapproved by what has existed in China and to some extent Indian economies. These economies had embarked on autarky policies but experienced stagnant growth. They eventually decided to open their economies. China opened its economy in 1978 while India opened hers after 1990. Some economies such as Taiwan and South Korea that have emphasized to exporting to developing countries have grown very strong.

2.2.4 The neo-classical, free market counterrevolution.

These neo-classic theorists put their emphasis on the role played by free markets, open economies and privatization of the inefficient enterprises. They do not attribute underdevelopment to external and internal forces as laid out by theorists. They attribute underdevelopment to too much government intervention in the economy.

According to the neoclassic counterrevolution, underdevelopment results due to poor resource allocation, incorrect pricing and too much government interference. The neoliberals argue that permitting competitive markets to flourish, privatizing state owned enterprises, promoting free trade and export expansion allowing in investors from developed countries and reducing government regulations and price distortions in factor, product and financial markets. A new wave of economic mechanisms emerged to the economies of United Kingdom when Margaret Thatcher was the Prime Minister (1979-90) and the United States economy when the President was Ronald Reagan (1981-9). They agitated for a free market as a way forward for economies. Also the Mexican president, Lopez Portulo (1976-82) hosted dignitories from the North and South for a New International Economic Order (NIEO). This was meant to bring about a sustainable development where there was going to be a vision of revised global economy and the needs and aspirations of developing economies were to be given new and greater consideration.

A debt crisis resulted as a result of recession in the industrialized nations. This began in 1982 and it led into poor terms of trade especially of the products from developing countries.   African countries and Latin America had accumulated huge persistent debts externally which made these economies suffer enormous crises.

It is important to note that neoclassic approach developed by Solow, capital accumulation sustained adequate saving, increase in the working population and technical progress increases the productivity of factors and consequently a steady growth. It is however noted that the growth path is exogenously determined.   New growth theory seeks to provide an endogenous explanation of growth of the economy. Technical progress is able to generate the growth process.

Governments use these theory basic concepts to do their roles towards the welfare of the nationals. The role of government in economic development includes the following;

Provision of social economic infrastructure such as schools, hospitals, roads, energy source and many other facilities so as to enhance economic productivity. This will ensure an educated, skilled and health population that is able to create wealth that is responsible for economic growth.

Government also has a noble responsibility of nursing and stabilization of the economy. This is useful to regulate the level and stability of the economy. Tools such as monetary and fiscal policies help government to influence level of prices, output, employment, investment and balance of payments.

Governments have a role of preserving and developing their natural resources in a sustainable way. This involves discovery of new resources how they are harnessed and taped with a sole responsibility of bringing about economic growth and development. This helps to improve the country's trade, earn foreign exchange that will ultimately elevate the way of life of the citizens.

Ensuring an equitable distribution of income and is one of the great roles that governments are supposed to play. This can be done through the taxation policy and maximizing social welfare so that even the grass root people enjoy a descent way of living. This requires a good governance and promotion of fundamental human rights where every citizen enjoys his rights that include promotion of social justice. Good policies should be able to change people's attitude that endangers development and so that they are able to develop globalization ideas of competition and elimination of racial discrimination. A culture of saving and investment need to be inculcated to the masses.

In his theory, Rostow explained economic growth in a historical perspective in different stages where capital accumulation is a driving force. These stages include; the traditional society stage, the pre- condition stage, the take off stage, the drive to maturity stage and the high mass consumption stage. These stages are in the range of subsistence to modern industrial economies.

Most economies in the developing world have consistently stayed in the precondition stage where there is low productivity that does not fetch enough foreign exchange for the country's import needs. There is a lot of import expenditure in relation to the export receipts. On the other hand some economies are growing so fast than Rostow anticipated. Some of these include the G8 economies.

There is need for the growing economies to continue adopting new ideas so that they are able to participate in global affairs. It involves adoption of new ideas, skills and knowledge which are a pivot to economic development. There increased transfer of technology and increased use of other co operant factors of production that result into improved efficiency in domestic industries. Technological innovativeness is encouraged and this multiplies the growth of economic activity.

Chapter 3

3.0 BASIC CONCEPTS IN PLANNING FOR DEVELOPMENT.

Implementation of development plans were not given due attention until 1980's when a few of these countries began to hinge on planning for their economies. Since then they have come to realize the importance of formulating and implementing national development plans.

Planning has become a mainstream of development. Government ministries usually draw plans for 5 years (a 5 year development plan). A comprehensive development policy framework can play an important role in accelerating growth and reducing poverty. For example the current World Bank comprehensive development frame work which requires countries to come up with Poverty Reduction Strategy papers in order to get financial and technical assistance.

Economic planning may be described as a deliberate government attempt to coordinate economic decision making over a long run and to influence direct and in some cases control the level and growth of a nation's principle economic variables. For example income, consumption, employment, investment, savings, exports, imports, among others to achieve pre determined set of objectives.

Economic development plan is simply a specific set of quantitative economic targets to be reached in a specific period of time with a stated strategy for achieving those targets. Economic plans may be comprehensive or partial

- A comprehensive plan sets its targets to cover all aspects of the national economy.
- A partial plan covers only part of a national economy. It might be agriculture, industry, public sector, and many of such sectors in the economy.
  There is need to formulate government objectives by the planning authority. The objectives are formulated basing on the general problems facing society. For example accessibility of clean water, health and education, opening of the upcountry roads and maintaining the existing ones.   Planning is done in consultation with all sectors of the economy such as agriculture, energy, environment, information and technology. The plan formulation involves both the top- bottom and bottom –up approaches. In Uganda the central Planning Authority is solely responsible to draw the national plans and co-opting the plans from local authorities for financing by the Ministry of Finance and Planning. The plans are adopted by parliament by confirming them or making the necessary adjustments.

The planning process can itself be described as an exercise in which government first chooses social objectives, then sets to various targets and finally organizes a frame work for implementing.

Supporters for economic planning in developing countries argued that the un controlled market economy can often does subject to those nations to economic dualism. (Traditional and modern), fluctuating prices, unstable markets, and low levels of employment. In particular, it is claimed that the market economy is not geared to the principle operational task of poor countries, mobilizing limited resources in a way that bring about structural change necessary to stimulate a sustained and balanced growth of the entire economy. Therefore planning came to be accepted as an essential and pivotal means of guiding and accelerating growth in almost all developing countries.

Another important ingredient of planning is monitoring and supervision. Many times good plans are made and funds reimbursed but implementation is not done. Funds are at times misallocated to other projects or swindled. And at times sub- standard work is done. Supervision and monitoring requires an independent body to oversee what is being done. A monitoring plan is drawn and this should be quarterly, semi-annual and annual evaluation plans.

In developing countries, there is at times no through monitoring. People are so dubious. They want to get rich quick especially countries like Uganda. The cause of this is most likely the history of this country where some of the past leaders were the victims of the corruption scandals. There is an African adage that says " a breast feeding mother should not go to steal, otherwise the child will also be a thief".

It is however important to note that development planning is constrained by lack of reliable statistics and information. In our communities there is no culture of keeping and recording information due the prevailing levels of illiteracy and also failing to release information. There is also existing oppositions for the ruling governments in these countries. These become contradicting forces that make plan implementation very difficult. This is on top of poor monitoring that is coupled with corruption and all this constrains development planning.

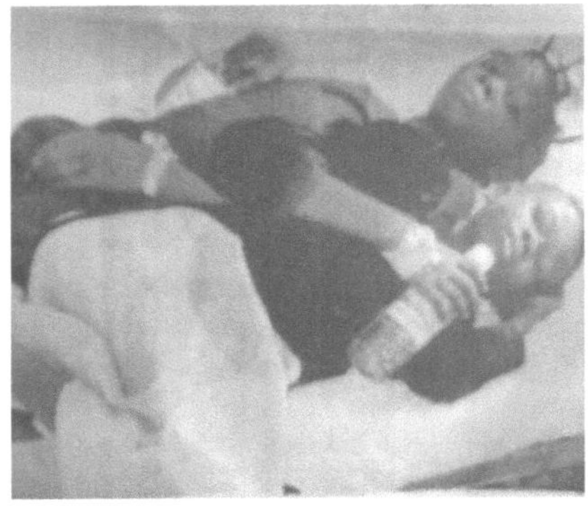

Photo 1 (News paper).   Photo. 2 (by researcher)

In photograph1, there is an evidence of poor policy and planning where the health sector in Uganda lacks the required support by government, non-governmental organizations, the civil society and the politicians. It was taken at Arua referral hospital in northern Uganda. Patients share beds in the hospital. The acting Principle Nursing of Sr. Sally Obiru, said sharing beds exposes patients to contracting diseases from each other. This is in addition to shortage of drugs and inadequate medical support staff in health centers. This calls for a strong policy on health care of the population.

Photograph 2,(taken on 29th Nov 2011) is a clear indication that there is poor policy planning and implementation on harvesting of forest products in Uganda. The photograph was taken from a certain household in Kabale municipality.   There is harvesting of young trees. These are in form of sticks instead of big logs and this poses threat to the environment. 275 sticks were counted weighing an average of 40kg. Comparing this with a selected log of a tree that had grown for 10 years and was weighing 165kg (from one tree), it shows a big loss (45375kg) of wood material for the just harvested material in the photograph. It is clear that pre mature harvesting leads to loss of wood material a situation that does not lead to sustainable development.

Chapter Four

4.0 THE RATIONALE FOR DEVELOPMENT PLANNING.

The early wide spreading as a development tool is rested on a number of fundamental economic and institutional arguments.

4.1 Market failure

Markets in developing countries are characterized by imperfections of structure and operation of commodity and factor markets. Markets are often poorly organized and there is existence of distorted prices which often scares producers and consumers that are responding to economic signals. It is therefore argued that governments have an important role to play in integrating markets and modifying prices. Market failure further leads to gross disparities between social and private valuations of alternate investment projects.

In the absence of government intervention, therefore the market is said to lead to misallocation of the present and future resources or at least to an allocation that may not be the best in the long run social interests.

The market failure argument is perhaps the most open quoted reason for the expended role of government in developing countries.

4.2 Resource mobilization and allocation.

Developing countries can not afford to waste their limited financial and skilled human resource on unproductive ventures.

Investment projects must not be chosen only on the basis of partial productivity analysis dictated by individual capital output ratios but also in the context of an overall development program that takes into account of external economies, indirect negative/positive results and long run objectives.

Skilled workers must be employed where their contribution will be mostly widely felt.

Economic planning is assumed to help by recognizing the existence of particular constraints and by choosing and coordinating investment projects so as to channel scarce resources into their most productive outlets.

In contrast, it is argued, competitive markets will tend to generate less investment and to direct that investment into areas of low social priority for example consumption of goods of the rich.

4.3 Attitudinal and psychological impact.

It is often assumed that a detailed statement of national economic and social objectives in the form of a specific development plan can have an impact on the attitudes on a diverse and fragmented population.

It may succeed in mobilizing people behind government in a national campaign to eliminate poverty, ignorance and disease.

By mobilizing popular support and cutting across racial, religious or tribal factors with the request of all citizens to work together towards building the nation. It is argued that an enlightened central government through its economic plan, can best provide the needed incentives to overcome opposing inhibiting forces and often diverse forces from different sections and traditions in a common goal for widespread material and social progress.

In view of the above, good development plans do not reflect the realities on the ground for example;

- Achieving high per capita income has not necessarily meant an improved welfare of the citizens as far as developing countries are concerned. There are characterized by corruption scandals where government funds meant to cater for the population ends up being swindled or misappropriated. There is therefore increased child mortality and maternity rates, shortage of drugs in hospitals, people dyeing of hunger, and general poor health of the population. There is widening income disparity between the very rich and the very poor.
- Despite good plans of accessing education to all, such programs end up being political. Politicians use such programs to achieve their ends. They do not lead the people very well to understand the essence and motive of such programs. They instead wish to gain political will. A case in point is Uganda where introduction of Universal Primary Education in 1996 led to general decline in standards of education. The political announcing of the program was political motivated. It excused the community to participate towards the education of their children. Parents were made to understand that government was to take care of education of their children. What happens today is that some parents even refuse

to buy the scholastic materials for their children hoping to get full support from the government. The end result has been increased school drop outs.

- There is failure to increase international competitive capacity in developing countries. Even when developed economies have opened their markets for mutual trade, the productive capacity in developing countries is still very low. Subsistence production with poor production techniques still dominates. There is therefore a low rate of foreign exchange necessary for importation.

Table 1 GDP and merchandise trade by region, 2004-05
(Annual percentage at constant prices)

		GDP		EXPORTS		IMPORTS	
		_2004_	_2005_	_2004_	_2005_	_2004_	_2005_
	North America	4.1	3.4	8.0	6.0	10.5	6.5
	United States	4.2	3.5	8.5	7.0	11.0	5.5
	South and Central America [a]	6.8	4.9	12.5	10.0	18.5	14.0
	Europe	2.3	1.7	7.0	3.5	7.0	3.0
	European Union(25)	2.2	1.6	7.0	3.5	6.0	2.5
	Common wealth of Independent						
	States	8.0	6.6	13.0	4.5	16.0	16.5
	Africa and Middle East	5.7	4.5	7.0	7.5	13.5	12.0
	Asia	4.2	4.2	14.0	9.5	14.0	7.5
	China	10.1	9.9	24.0	25.0	21.5	11.5
	Japan [b]	2.3	2.8	10.5	1.0	7.0	2.5
	World	3.9	3.3	9.5	6.0	....	....

[a] Including the Caribbean.
[b] Trade volume data are based on Japan's customs statistics. National account data report a markedly stronger export and import growth in 2005.

Source: WTO

Series 1- GDP 2004   series 3- Exports 2004   Imports- 2004.

Series 2- GDP 2005   Series 4- Exports 2005   Imports- 2005.

From the table/ graph, North America's exports and imports improved by approximately 6%. This was the same rate as the world rate in 2005.North America's real merchandise exports and imports expanded by about 6 per cent, the same rate as world. It is reported that for the first time in eight years US merchandise exports rose faster than world exports. For Africa and Middle East, merchandise imports rose more than exports. This reflects fewer receipts from the exports compared to the payments on the imports which results in a balance of payments deficit.

This calls the concerned countries with negative balance of trade especially developing countries to put into place formidable policies to improve their trading positions. Their trend in the international trade is not healthy.

Liberalization of international trade has done a good job in economic and social transformation. It is however necessary for government controls to some extent to control the activities of the private sector. At times the private sector does the business using speculative tendencies which affect the price mechanism and ends up affecting the welfare of the people. For example, business speculation leads to back marketing and hoarding that at times develops into market imperfections. Full scale trade liberalization leads to unfair competition in the developing economies where infant firms have not yet developed the capacity and technology to compete with the established firms. In this case a mix of trade liberalization and government interference is recommended for proper growth of economies.

Programs to reduce population growth rates have not yet achieved the desired results. Population growth rates are still high. For the case of Uganda the population growth rate is still over 3%. The increasing population in these countries does not match with the population quality so as to engage in production. They end up being dependants on the few working classes and on respective governments. The major reason of failure to curb the high population growth rates is the existence of social-cultural factors. In most societies having many children is prestige in society. These children are regarded as source of defense and labor. The governments, civil society and politicians need to show a lot of commitment to change the status quo of our societies so that the attitude of the population towards certain things can change.

4.4 Foreign Aid.

The formation of detailed development plans has been necessary condition for the receipt / receiving of bilateral or multilateral foreign aid. With a number of projects/sector projects, governments are better equipped to solicit foreign assistance (financial and technical) and persuade donors that their money will be used in a well conceived and internally consistent plan of action. These are normally development programs to aid poor countries to overcome some social political challenges. For example the Inter-American Development Bank (IDB) that gives long-term development loans to Latin American countries, the African Development Bank (ADB) which gives loans to African countries to promote social-economic development in these countries. The European Development fund gives loans to European associates all over the world; in Africa, Caribbean and Pacific regions. The Poverty Reduction Strategy Plan under the comprehensive development framework under the IMF and World Bank requires governments in need of support to submit the Poverty Reduction Strategy Papers (PRSP'S). In Uganda's case, it was the first in 1997 to submit her poverty eradication action plan (PEAP) for financing and implementation. However, the Uganda government is designing a new national development plan that can form sectoral development programs, policies and projects.

Most of the developing countries need aid to cover up their budget deficits. This is because the taxable capacity in these countries is still low. The structure of their economies does not provide enough tax bases. Aid therefore helps them to expand investment that increases employment, fills the foreign exchange gap needed to purchase the essential imports like drugs.

It is however noted that many communities want to seek external assistance to address their problems. This to some extent deprives the communities the capacity to think for solutions to their challenges. This assistance should be rendered by first understanding the community context and social dynamics in this particular society. This helps to discover the community's aspirations and know what type of assistance to give. Once assistance is rendered, monitoring and evaluation programs should be drawn in collaboration with the beneficiaries to inculcate a sense of ownership and identity.

Planning should involve all stakeholders and these include the community, civil society organizations, government departments and public –private partnerships.

Civil society organizations/public/private partnerships of development.

These partnerships refer to the institutional and organizational partnership arrangements for improving the quality, provision, utilization and sustainability of socio-economic policies, programs, projects through a coordinated delivery of services within the structure of the economy.(agriculture, industry and services).

It focuses on significant local, national and international priorities. Examples of such partnerships at the international level include the efforts of national governments, civil society organizations and private sector organizations in the achievement of the millennium development goals.

Locally, such partnerships include local NGO's CBO's, FBO's and public sector for instance higher and local governments as well as medium, large and small business enterprises working together to achieve national development goals that fit into the achievement of MDG's.

4.5 The role of CSO,Private and Public partnerships in community development.

These partnerships promote participatory planning where by different organizations has an input to the national development strategy. Usually when national governments are planning they should include civil society organizations and private sector organizations. This ensures effective allocation of resources equitably to all sectors of the economy. Cases are noted in the fields of agriculture, industry and services.

These partnerships promote transparency and accountability given the role of civil society in checking government programs and how they are implemented. The private sector has an important role in generating economic growth for community development.

The role of these partnerships also includes sharing information and experiences for effective service delivery. As they share experiences, there is organizational capacity building, thus ensures flexibility and adaptive procedures which can facilitate accountability to the beneficiaries and mostly increasing efficiency and promoting development results.

Partnerships also promote joint research and development projects including needs assessment, feasibility and business planning as well as participatory monitoring and evaluation to identify critical gaps and development needs and services in any given community where these organizations are operating.

Partnerships in addition to efficient resource allocation provides a pool of resources in form of financial, technical, technological as well as human resources that are missing either in one organization and are needed in another organization.

Chapter Five

5.0 THEORY, POLICY AND INSTITUTIONS

The policy making process seeks to determine the correct policy instruments from;

The theory of socio-economic reality, for instance social, economic change of the existing situation. There are policies meant for economic development and these include;

> Policies meant to provide public services such as infrastructural development which is a cornerstone of investment. These include health, education, housing, power grid, roads. They enhance increased production and propels troubleshoots economic growth.
> Policies by government to meet broad economic objectives. Such as controlling inflation, reducing unemployment, and maintaining sustainable growth. Tools of monetary and fiscal policies and regulation of financial institutions are used.
> Job creation through industrialization, small business development, diversification of production, proper manpower planning and training.

In developing countries such measures have not successfully achieved the intended objectives due multiplicity of reasons.

Rural areas are still underdeveloped. There are poor services rendered in form of health, education, roads, energy needs, and housing. As a result, there is unhealthy, unskilled or semi skilled, population which limits capacity building towards production and this limits on the export base, creativity, innovativeness and entrepreneurship.

In addition there is a small size of the market due to low incomes. There is therefore little or no incentive to invest in such economies which limits the employment opportunities to the population.

Heavy reliance on agriculture that depends on nature. Nature is the mother of people in developing countries especially the tropics. Poor climatic conditions mean poverty, hunger, disease and death. This is due to poor technology in these countries. In addition agriculture is mostly at subsistence level where production is meant for household consumption. This contributes a meager share on the Gross Domestic Product, with no employment opportunities created.

Heavy debt burden in the poor countries that continue to service their huge debts fail to save. Heavily indebted countries fail to provide the population with the public services leading to the poor welfare. Poor economic indicators such as high poverty rates, low life expectancy, poor health, education and housing, high illiteracy rates and poor environment quality have remained the order of the day.

- Institutional capacity for policy making and implementation. For instance institutions to manage the policy making process. In other words, what is the capacity of government, business sector, civil society organization in managing the policy making process?

5.1 An overview of Development Policy Management.

What are the elements of policy management?

Development policy management embraces both policy design and policy implementation. Their key elements include;

- Vision of development for instance the general consensus of where we are going as an organization or community for example in the next 25 years for a long term vision.

- Goals and objectives of development. The objectives should be Specific, Measurable, Achievable, Realistic, and Time bound. These objectives should cover all the main sectors of the economy. These will aim at efficient allocation of scarce resources, reduce the rate of unemployment, reduce income inequalities, improve and strengthen the market mechanism with the sole aim of achieving sustainable growth and development.

Vision of development

This is what gives the system a direction. In what direction does a nation or community has to change reality towards a certain end. A vision shows what the community or nation wants to become. This vision of development depends on the concept of development.

Traditionally, development has been focused on economic growth for instance on theories of modernization of attaining the same levels of development as the United States, UK, Japan. But what is the meaning of being rich. We should also consider various dimensions of development such as social, political and economic.

Goals and objectives.

One aspect of development policy management is setting goals and objectives. A goal is a long term purpose of policy and this purpose must be related to the vision. The major goal of economic development is to improve the welfare of people in society. It involves provision of food, shelter education, and medical care as some of the indicators of the standard of living. Therefore the development process aims at increasing the welfare of the people through raising employment opportunities, increased incomes, availing basic needs and improving the medical care.

5.2 Instruments of policy.

These can be tools used to effect social economic change for the attainment of 3 objectives of economic growth and development in developing countries. Such instruments of policy include;

Laws and regulations. This implies that whenever there are laws and regulations on the environment. For example NEMA, Ministry of Water, Lands and environment

The policy frame work; which is a set of incentives and support services for the actors/ agencies in a given sector of the economy. This may include support to the households in order to reduce poverty through education (UPE, USE, higher education) as well as improving their health.

Institutions. These include NEMA,UIA,URA,UWA.

Strategic investment plans. These can be long term or medium term plans. These can be documented sets of action designed to develop a certain sector of the economy which include a strategy for allocating resources in order to maximize development efforts. It should involve how much resources should be put in different sectors like education, health, environment among others.

Development policy and management has five levels;

- The global level. This is where the main actors are the multi lateral and bilateral agencies. Such as UNDP, United Nations Development Program, the World Bank, IMF. These agencies belong to all nations. At the multi-lateral level, policy making has been globalised through multi-lateral agencies of the UN and the G8 in which development initiatives supplement national initiatives. At the international level, there NGO's such as OXFARM, Action Aid World Vision Compassion.
- At the national level there are 22 organizations such as MTN, Coca-Cola among others. In Uganda, National Planning Authority is responsible for coordinating, planning for the economy as well as creating partnerships within ministries, civil society and private sector organizations. The bargaining power is both through

political administrative and participatory approaches especially through planning units of ministries and other relevant parliamentary committees.

- The sector level these may include the sector and the sub sector level. For example development plans for secondary education, primary education, and higher education. Most activities at the sector and sub sector level attempt to build capacity for achieving the bigger sector objectives.
- The local government. Local governments are concerned with managing development programs through approaches like decentralization as they contribute to the realization of the national development programs. The local governance system allows for local planning right from the village up to the national level. As part of the change from a state oriented to market economy, decentralization was introduced in some countries. In China following the 1949 revolution, the economy was divided into small production units from the village commune right to the large scale enterprises, but the central government remained the supreme in decision making and control. The central government remained with the authority of service provision and could ensure that health care was accessible and affordable to the grass root population.
- The community based level of policy. This is a level that is always neglected and is supposed to promote participatory planning using a bottom up approach that should involve civil society organizations, private sector in order to contribute to the central governments national plans.

5.3 Instruments of policy for Environment Conservation.

1- International laws on certain issues. For example laws on endangered species such as gorillas, white rhinos, birds, wetland management. These are able to protect the environment. In Uganda some habitats are legally protected because of globally shared resources. For example the birds that fly from Europe to Africa. International protocols. For example the Kyoto protocol that govern the law of the ozone layer.
2- Global partnerships for sustainable development which involves building institutions that works together among countries where necessary because destroying the ozone layer affects the public.
3- Economic policy instruments: these include macro economic policies that are sensitive to environmental protection; there can be mechanisms on the environmental impact assessment to assess the consequences.
4- Introducing taxes on organizations that deplete the environment and also water treatment points.

5- Decentralized environmental governance through the establishment of local environment committees which make bye laws. Under urban decentralization, urban planning is essential in developing as it was done in developed cities. Due to lack of planning cities are poorly planned, for a long time and the private sector haphazardly due their allocation of resources in urban centers. By the time the government authorities try to intervene by having planned cities a lot of damage is made to assets worth of millions of dollars. Of recent, Kampala Authority realized there was unfair development of Kampala city. In a bid to rectify this fake trend the city authorities decided to implement the City plan. They have demolished property worth millions of dollars by pulling down the buildings in the city. Most Ugandans are asking, where were the authorities when private investors were electing these assets. A similar will in future happen in the small towns of Uganda since the authorities are not bothered to implement the urban plans of these towns. Where implementation is being done it is marred by corruption. This is just unfair and it calls for proper urban planning and implementation. Growth cannot wait for slow actors. It will always take place with or without planning even though it will be inequitable. There has been a lot of urban planning around the world where developing countries can borrow a leaf.

There were urban planning initiatives for urban planning in Britain way back before the World War 1. In 1910, a conference was held in London that constituted the Royal institute of the British architects, the London Society and the Town Planning for the future of the London city. Regional Imagination, central to the later development of planning, was clearly visible in the program of work undertaken during the First World War and the associational networks were important part of early professionalization in planning in Britain (Hewitt, Lucky, 2011). In most developing countries, governments do not plan the cities. It is the initiative of the local population to initiate urbanization along the main ways. The people reside along the roads for commerce and easy accessibility. This is like a natural activity that has no prior planning. A collection of many families will create a market demand for small shops that will eventually attract tax collectors. This is how the authorities come in when there is unplanned allocation of settlements, factories and roads. Correcting the errors made by the natural process costs the private sector a great deal. There are many up many coming trading centers along the roads and the government authorities are not bothered about their development. But one day the government authorities will come up to rectify the wrong planning and this will cause frustration. Planning of urban centers should be done by the civil society, the private sector in collaboration with the community. It should not be the centers to grow up by themselves. One of the old man James testifies that he built the first house in Muhanga and other residents continued joining him until the present day. Muhanga has just gained a town Board status with a town clerk and other town leaders. Of course there is no proper planning for this town. Another well elaborated example was during the 1970's when United States tried to curb urban disinvestment. In the same period, Milwaukee and Wisconsin implemented urban planning programs. They conducted Relative Residential Status (RRS), and Preservation Planning meant for the growth of the US cities and its neighborhoods.

Photo 3(by researcher)

The photograph was taken on the 30th Nov. by the researcher. It shows growing Kyanamira trading centre 6 km from Kabale municipality. The centre is attracting a growing population from the rural areas. People are building temporally structures with no planning initiated by government. As other centers have always grown this one will grow too but the challenge will be to put the already existing on an urban plan. Another important feature observed is construction of the structures in the road reserve which is supposed to be 15 meters. Construction of the houses is being done in a distance of 6-10 meters from the road. Urban planning of this area will be costly because it will involve pulling down structures meaning loss of resources.

5.4. Policies for economic growth and development.

Economic growth means the qualitative increase in production as measured by GDP/GNP. It leads to increased production of goods and services of any country or society. Economic development means economic growth plus structural change which all lead to the improvement in the quality of people's lives.

The concept of broad based sustainable economic growth is important since all the sectors should share the growth and enjoy that part of growth in order to promote economic development (improving the quality of people's lives).

In developing countries, we should conserve and utilize the natural resources without depleting or destroying these resources because it would be a source of living for the next generation.

There is need to improve on the public services by the respective governments. This therefore calls for proper planning and policy to mobilize resources and effective allocation of these resources for maximum benefit. Failure for proper policy for economic growth and development will lead to the following;

Exhaustion of natural resources and degrading the environment that will directly impact the welfare of the people across continents. There has been encroachment on the forests and swamps due to economic activity for economic gains. It also leads to social costs such as water and air pollution, due to no proper policies of industrialization. Congestion and traffic jams due to lack of adequate urban planning. There is need for proper resource utilization and preservation for sustainable development. This can easily be achieved through proper policy and planning.

Unplanned economic growth ends up creating a wide gap between the very rich and the very poor. Many people cannot afford the basic needs. As a result many associated problems will arise. These include unemployment, social tension, inadequate demand and social tension. There is always need for the youth to move from rural to urban areas to look for jobs as has been the case for developing countries. Poor quality services will be enjoyed by the majority urban slum dwellers due in adequate facilities to a huge population in these areas. There are also inadequate domestic markets due the wide income inequality in the developing economies which scares away domestic and foreign investors.

There is an observation that there is a lot of extravagancy among the population in developing countries especially in African countries. This is evidenced by the expensive life they lead at the expense of capital accumulation. They hold expensive parties, expensive burial ceremonies and exaggerated government expenditures. The marginal propensity to save is still very low. In urban cities in Uganda, there is a lot of expenditure over the weekend and evenings after work especially in Kampala. This is at the expense of saving for investment.

Economic involves so much hard work. In our societies most of the people have not yet realized the importance of hard work especially men. Women caterer for the families while men are at the drinking places morning to evening. This deters development since men in most societies are un exploited resource. Men need special programs to wake them out of slumber if desired growth is to be achieved.

High population growth in developing countries needs to be planned for if these countries are to achieve required economic growth. The high population growth rates lead to a declining per capita and increased dependants hence retarded growth. It creates misery, poor nutrition, poor housing and general deterioration of the welfare.

Another important ingredient needed for economic growth and development is improved skilled human resource development. Proper planning involves upgrading the education system so as to address shortages of professional, scientific and managerial manpower in developing economies. Uganda like most developing countries faces a problem of shortage of skilled man power which has negatively affected the development of the economy. For instance, Uganda's education system is characterized by the curriculum that is theoretical and does not address the needs of the community. It instead creates a problem of school leavers who are job seekers instead of job makers who end up being unemployed and yet the country lacks man power in certain critical areas such as mining engineers, food processing and many other sectors of paramount necessity to community needs. Proper policy and planning for education programs would eliminate most of the problems most of these countries are facing such as diseases, famine and hunger, poverty, social cultural and poor attitudes that impacts on growth, poor technology and many man's common problems.

At school we have school rules and regulations that we use to govern the school. There is school council where issues that concern them are discussed and solutions are sought. Among the school policies we have a policy of each of the student to belong to one of the clubs. These clubs impart virtues that are not delivered in the classroom. These clubs include; the wild life club, straight talk club, scouts club, Red Cross club, Purity club, Rotaract club, Patriotism club and many others. In these clubs students are able to learn important issues that surround them such as fundamental human rights, pertinent skills and lifestyles.

Students' attitudes are able to change through the programs that are designed in these clubs.

5.5. Results of proper policy management to the population.

There will be increased material prosperity. These results into goods for consumption, capital goods and infrastructural development that will enhance increased production.

It leads to improved welfare of the population and specifically to the common man or the low income earners. This entails accessibility to medical care, water, education, food and shelter.

It improves the political, social and economic conditions of a nation or region. This includes good governance and social justice, increased employment opportunities, improved political stability that generally creates harmony in society.

There will be improvement in the balance of payments position. This will be as a result of improvement in production techniques that will lead to improved production for exports. Value addition through processing leads to better products that can fetch more revenues and are also able to compete on the world market. There will be increased government revenues through taxes which will be a basis for providing public services.

6.0 Recommendations.

Transformation of poorly agricultural subsistence communities of developing countries to modern economies that are industrialized. there is a lot to learn from giant economies where there was massive injections of capital and re arrangements of economic structures. For instance the US and the European economies.

Reducing the gap between superior and inferior elements in society should be emphasized. This is as a result of dualistic tendencies where there is co-existence of desirable and undesirable practices for example co-existence of traditional and modern sectors in the same economy.

There is need for the regulated private sector to eliminate poverty, provide employment, reduce illiteracy, and improve the health status of the population. This can easily be done through policy and planning of respective economies. There is need for domestic and international reforms that are meant to improve economic performance through public and private economic involvement. The only challenge is determining the extent to which the government and private sector should take on the responsibility. Numerical proportion of this responsibility can not be easily ascertained.

For effective planning, governments should put emphasis on decentralizing powers to the local levels for effectiveness and efficiency. This will enable the local communities and local authorities to take part in decision making which creates a sense of ownership for sustainable development. For instance, decentralization has been a majorly making process in the North and South. Federal States are always empowered to make significant decisions such as Mexico. If developing countries are to attain substantial economic growth rates, grass root planning is essential. The common practice is a lot of powers vested in politicians and the executive arm of government to do resource allocation in developing countries. This does not yield expected results.

Developing countries' governments should respect fundamental human rights. The welfare of citizens needs a lot of attention by the planners and policy implementers. A lot of irresponsibility is noted by the civil service, politicians and all those entrusted with powers in governments. For example in Uganda, cases of government negligence is noted in areas of ; health where AIDS patients fail to access drugs, suffering of the marginalized such as orphans on the urban streets, mud persons sleep on the streets, the starving elderly, high child mortality and maternity rates, among other problems faced by the grass root population.

Governments should try to reduce and minimize the public debt. There is a negative effect on growth, distribution, wealth, and capacity building. Debt servicing causes capital outflow especially when the debt was contracted to fund unproductive projects. An external debt directly affects other macro economic variables such as capacity utilization, and growth, terms of trade and wealth creation to the heavily indebted countries.

There is need for proper man power planning and training in developing countries. This would increase the skilled man power and entrepreneurship. It needs an improvement in education policy. For example change of the curriculum from the theoretical approach to a practical and self discovery androgenic approach. The system of education has a lot to change so as to meet the needs of the local population. It needs to impart skills and required values for the good of the community.

There should be proper resource distribution of resources by government among the sectors, regions to enable equitable development. For example massive investment in priority sectors because these sectors will act as engine of growth. It also involves encouragement of setting of industries in the rural areas to curb rural urban migration.

Emphasis by government on massive investment of resources in infrastructure, agriculture and industrialization is needed. This promotes rapid industrialization where more employment opportunities are created. It also makes use of backward and forward linkages that is necessary for a self reliance economy. It also explores the importance of the comparative advantage of the respective economies which will improve on their trading positions.

Adoption of appropriate technology, technology transfer from across countries. It will be helpful in improving the quality output for international market. The quality of output from developing countries is not good enough. It is either semi- processed, unprocessed, or poor quality manufactured goods that does not guarantee better terms of trade. Improved technology is also useful to exploit the idle resources. People in these countries are in abject poverty yet there are a lot of idle resources that are not exploited. There is however an observation that up coming small scale technicians and entrepreneurs lack government support. Governments do not regard these as important towards growth. They are instead taxed for government revenues. Government effort to uplift indigenous technologies sand witched with the imported technology will bring about economic growth. It will initiate creativity and innovativeness.

There should be proper policy and planning to minimize the income gap between the very poor and the very rich. This can be done through upgrading education for all where the rural schools where the majority poor school children are well facilitated. Also emphasizes on income redistribution policy through progressive taxation where the rich are taxed more heavily and the tax revenue is used to provide public services for the entire population.

In 2009, world trade was dominated by worst financial and economic crisis where global output shrank. It is important to note that developing economies were hit worst. For instance Brazil, India and China exports dropped between a fifth and a third in the second half of 2008. The international community has tried all measures to see this situation change. There is however other challenges that are not being addressed by respective governments and are affecting the global economy such as the climate change. Climate change is one of the major aspects that all governments must address for economies to achieve sustainable development.

Chapter Seven

7.0 Conclusion.

Sustainability of growth requires proper mobilization of resources, legitimacy and powers of coordination so that these resources can maximally yield for the good of the nation and global social economic complex systems. Normative and analytical planning has not seriously put into account environmental sustainability and other pertinent issues such as land use and urban planning. Issues that have always been looked at to be affecting nations have always turned out to be challenges to the international community. These include poverty, disease, wars, environmental sustainability, governance, climatic changes among others. There need serous attention by everyone on the globe regardless of color, race, religion and where one lives. Proper policy and planning has not been done in most economies and this has hindered economic growth and development which has a direct impact on people's welfare, trade position and general social economic status of the nation.

Use of all stakeholders is a sustainable approach to provide solutions to people's concerns. Fragmented decision making processes deprives the community the capacity to conceptualize their needs visa-vies the solutions. A collaborative policy of all the actors is necessary right from the policy review, through the planning process up to implementation. While designing policy, rules, and regulations all interest groups should be consulted for an effective collaborative approach.

Man is always innovative so long as there are rewards attached. What is important is the motivation behind what takes place in communities. On top of basic skills imparted, rewards for new innovations creates the capacity for science and technology to take root, and this is a pre requisite to economic growth and development. Development theory and policy are very important tools for planning. It generates economic growth and development that forms a basis for better standards of living of the people.

References.

Cypher, James.M. (2009). *Process of Economic Development.* (3rd ed.). Taylor and Francis Routledge.

Gary, Paul, Green. (2010). *Community assets : Building the capacity for development.* Temple University Press.

Daily monitor Uganda News paper (1st Dec. 20110).

Hardcastle, David.A.(2010). *Theories and Skills of social workers.* Oxford University Press.

Hewitt Lucy. (2011). Towards greater urban geography: Regional planning and association networks in London during the early twentieth century. Planning perspectives 26, no.4 551-568.

Innes, Judith. E. (2010). *Community development, Urban political planning, Social participation.* Taylor & Francis Routledge.

John Dhumba Sentamu (2004). *Basic Economics for Eat Africa: Concepts Analysis and Applications.* Kampala : Fountain Publishers.

Kenny, Micheal. (2002). Planning Sustainability. In environmental politics, Taylor & Francis Routledge.

Ramirez, Rafael. (2010). *Business Planning for Turbulent times: New methods for Applying Scenarios.* Earthscan.

Salvadori, Neri.(2005). *Innovation, Unemployment and Policy in the theoris of growth and distribution.* Edward Elgar Publishing House Inc.

Setterfield, Mark. (2010). *Hand book of Alternative Theories of Economic Growth.* Edward Elgar Publishing, Inc.

Schimit, Deanna. (2011). Urban triage. Saving the savable neighbourhoods in Milwakee. Planning perspectives 26, no.4 569-589.

Stark, Andrew.(2010). Drawing the line. Public and Private America. Brookings

Institution Press.

Willis, Katie.(2011). *Theories and Practices of Development.* (2nd ed.) Taylor and Francis Routledge.